If You Lived Beside the DESERT

Gloria G. Adams

Slanted Ink
Stow, OH 44224

If You Lived Beside the DESERT

Gloria G. Adams

ISBN: 978-1-7367688-8-4

Based on a previous title by the author,
My Backyard Is the Desert

Images: Pixabay.com

Slanted Ink

www.slantedink.com

Stow, OH 44224

For
Pauly, Abby,
Elisabeth,
Maddy,
and Hannah

If you lived beside the desert, you might

SEE...

bright cactus blooms in springtime,

Cactus plants grow well in hot deserts because they are able to store water. They grow in deserts all over the world.

a solitary scorpion
scuttling across the sand,

Scorpions can have as many as 30 baby scorpions
They carry the babies on their backs
for the first three weeks after birth.

or even prickly cactus arms reaching into the sky.

Saguaro cactus can live as long as 200 years. They grow in in the Sonora Desert of Arizona and are the largest cactus in the United States.

If you lived beside
the desert, you might
HEAR...

the harsh screech of
a sky-soaring hawk,

Harris Hawks often use a saguaro cactus for their nests. They are one of the few types of hawks that hunt in groups.

the bark of a mama kit fox
protecting her cubs,

The large ears of desert kit foxes help them hear well. Furry footpads protect their feet from the hot sand.

or even the *rattle-hiss* yell of a burrowing owl.

Burrowing owls build their homes underground. They scare predators away by making a sound like a rattlesnake.

If you lived beside the desert, you might **SMELL...**

the sharp, spicy-clean fragrance of sage brush,

Sagebrush is an important plant.
It provides food for many animals,
including elk, jackrabbits, and sage grouse.

the fresh, woodsy spikes of a hardy agave plant,

Agave plants have strong fibers inside their leaves. These fibers can be used to make nets, ropes, sandals, and nets.

or even the honey-sweet scent of the acacia tree blooms.

Acacia trees have long taproots, which help them pull water from deep underground. This tree covers large areas of Australia.

If you lived beside the desert, you might

TASTE...

prickly pear cactus fruit as sweet as a melon,

Prickly Pear fruit is used to make candy and jam. It has other names, too, such as "cactus fig" and "tuna."

a crunchy snack of almonds plucked from a branch,

Bees are very important for almond trees. If none of the flowers are pollinated in the spring, the tree will not produce any almonds.

or even a bowl of rich, chewy dates.

The word "date" comes from the Greek word for "finger." There are over 2,000 different kinds of dates.

If you lived beside the desert, you might TOUCH...

a handful of
gritty, golden sand,

Sand dunes, formed by wind, can reach heights of more than 500 feet. Many dinosaur bones have been found in the Sahara

the swirly ridges of desert rock shapes,

Canyons form over millions of years when river water cuts through and wears away rock. This process is called "downcutting."

or even the delicate blooms of a fairy duster plant.

The fairy duster plant provides food for deer, as well as attracting pollinators like bees, hummingbirds, and butterflies.

And when the day ends,
you might watch the sky
fill up with stars,

if you lived beside the desert.

FUN FACT!

The dark patches around a meerkat's eyes help reduce the glare from the desert sun.

Polar Deserts are the largest deserts on Earth.

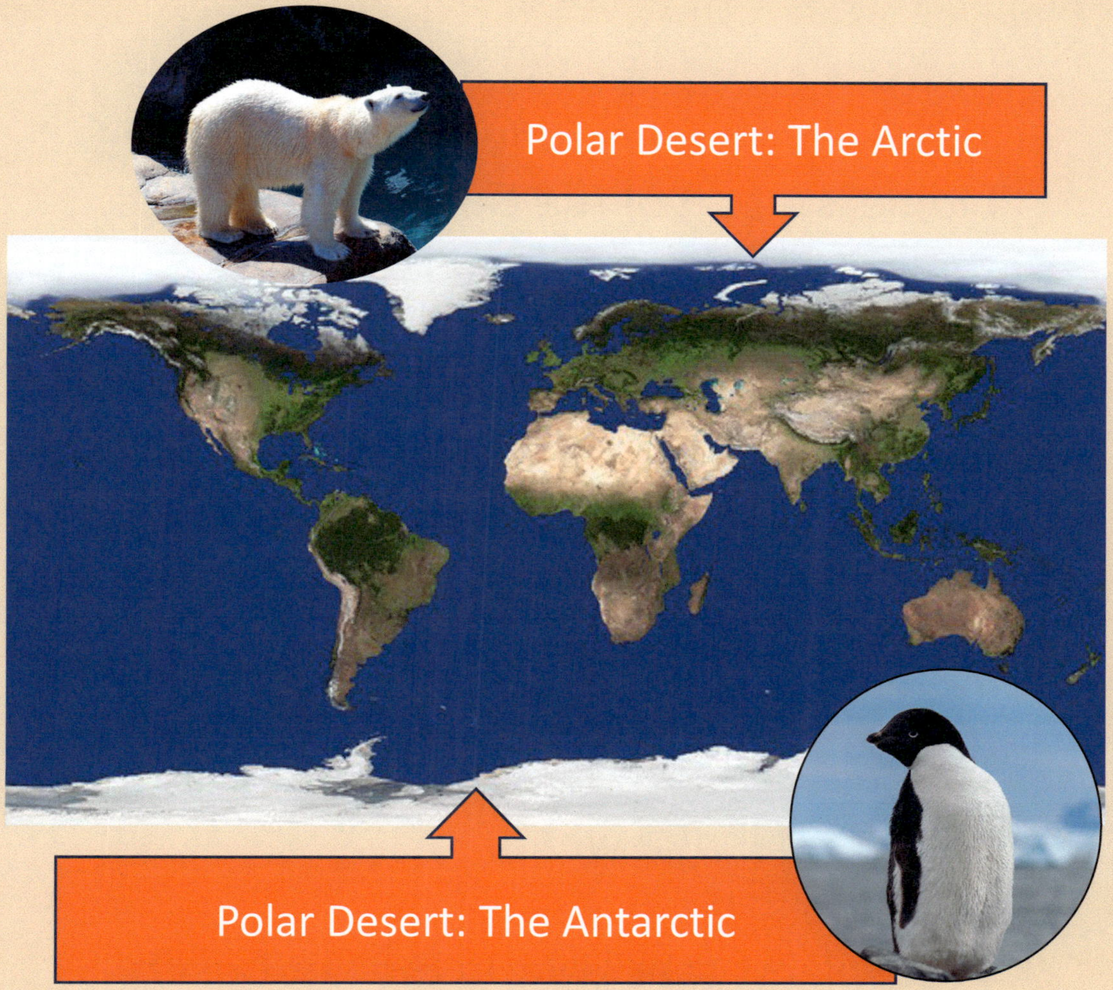

Where on the Earth do YOU live? What do you see, hear, smell, taste and touch in YOUR world?

Did You Know?

- The Desert is one of the Earth's **Biomes**. A biome is a part of the Earth that has the same kinds of plants, animals, insects, geology, and climate.

- While all deserts are very dry, not all of them are hot. Some deserts are cold, like the Arctic and Antarctic. They are called "polar deserts."

- The Sahara, in North Africa, is the largest hot desert in the world.

- Some plants that grow In the hot **Desert Biome** include many types of cactus, aloe, creosote bushes, and Joshua Trees.

Bananaquit Bird

- The plants and animals that can live in the desert need very little water. Camels can go for 15 days without any water. The Kangaroo Rat gets all of its water from the food it eats, so it never needs to drink any water at all.

- Lizards, snakes, rodents, meerkats, camels, coyotes, cougars, hawks, owls, tortoises, and scorpions live in hot deserts. Polar bears, arctic foxes, and penguins inhabit polar deserts.

- The birds, plants and animals in this book are all part of the **Desert Biome**.

About the Author

I am a former children's librarian and storyteller. I believe that reading to children and helping them become strong readers is one of the most important things we can do. Visit my website at www.gloriagadams.com.

~ Gloria G. Adams

Other books by the Author

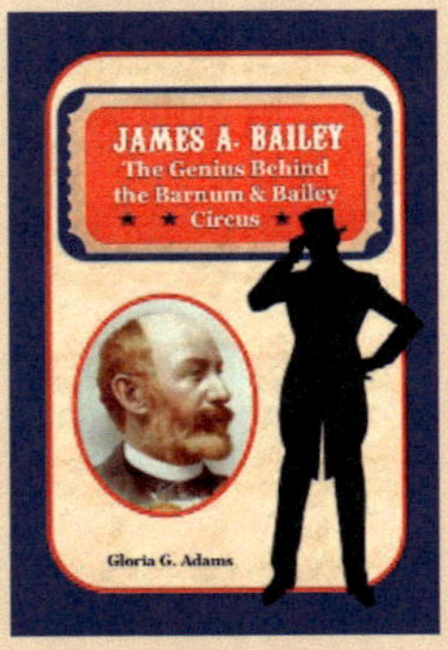

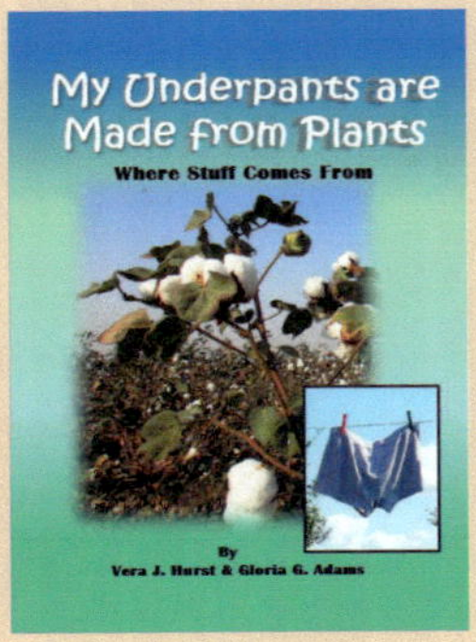

Made in the USA
Las Vegas, NV
07 November 2026

57672513R00021